T0164413

The Little Seal

The Little Seal

AN ALASKA ADVENTURE

Written and Illustrated by Ram Papish

SNOWY OWL BOOKS
an imprint of the University of Alaska Press

University of Alaska Press
P.O. Box 756240
Fairbanks, AK 99775-6240

ISBN 978-1-60223-068-2

Library of Congress Cataloging-in-Publication Data

Papish, Ramiel.
The little seal : an Alaska adventure / written and illustrated by Ram Papish.
p. cm.
Summary: A young northern fur seal observes huge bull seals as they choose territories and select mates, waiting for his turn to claim a spot on the rocky Alaskan shore.
ISBN 978-1-60223-068-2 (hardcover : alk. paper)
1. Northern fur seal—Juvenile fiction. [1. Northern fur seal—Fiction. 2. Seals (Animals)—Fiction. 3. Alaska—Fiction.] I. Title.
PZ10.3.P22245Liw 2009
[E]—dc22
2009017242

This publication was printed on acid-free paper that meets the minimum requirements for ANSI / NISO Z39.48-1992 (R2002) (Permanence of Paper for Printed Library Materials).

All illustrations by Ram Papish
Design and layout by Paula Elmes, ImageCraft Publications & Design

Printed in China

In memory of my grandfather, Don W. Fawcett, who inspired my love of nature. And with thanks to Karin Holser, who is a true champion of Alaska's fur seals.

About the Author and Illustrator

Ram (pronounced "Rom") Papish graduated from the University of Oregon in 1995 with degrees in art and biology and has worked as a field biologist all over the Western Hemisphere. He has studied nesting seabirds on several remote islands in Alaska and draws upon his experiences as a biologist to produce beautiful and accurate wildlife paintings.

The lives of seals are lives of the ocean, woven within its constant motion. With no earth or land within his sights, the little seal has spent 600 nights. But now he is drawn to the rocky land, to the place where his life began.

Smooth dark
forms splash and swirl as other
seals dash and whirl. Three seals burst from sea to air,
and plunge back down without a care.

The speeding seals move so fast, it's a wonder their pace can last. Soon they'll reach the shifting shore. Soon they'll be at sea no more.

Quick nip, and flee, and follow, three seals slide through waves that swallow. The hard wet stones are strange to feel after the seagoing life of a seal. Salt spray calls the three to stay, calls them as they move away.

Three seals scramble up the rocky beach, away from the sea and its watery reach. Now there enters into the story a big beach-master in all his fat glory. The mighty seal claims a spot to defend. No other male seal will now be his friend. Any that enter his rocky beach zone, in search of a spot to call his own, will face an attack both rapid and hot. All will flee before his onslaught.

Now the dark beast, so big and stout, gives a roar and turns about. The three small seals turn and run, two toward the sea and inland goes one. The first two turn and stare; the third seal's vanished to a grassy nowhere.

The little fur seal is a Sub-Adult Male, just three feet from nose to tail. Since saying Sub-Adult Male is a bit of a cram, from now on he's just called S.A.M.

SAM swims with the other young seal to a place males go to rest and heal. The young seals haul out to play and rest until they're ready for the big-seal test. Some seals bark and fight with vigor. They'll be ready when they're bigger. To get a spot on the breeding ground takes the toughest, roughest seals around, and only those who go through this bother will have a chance to become a father.

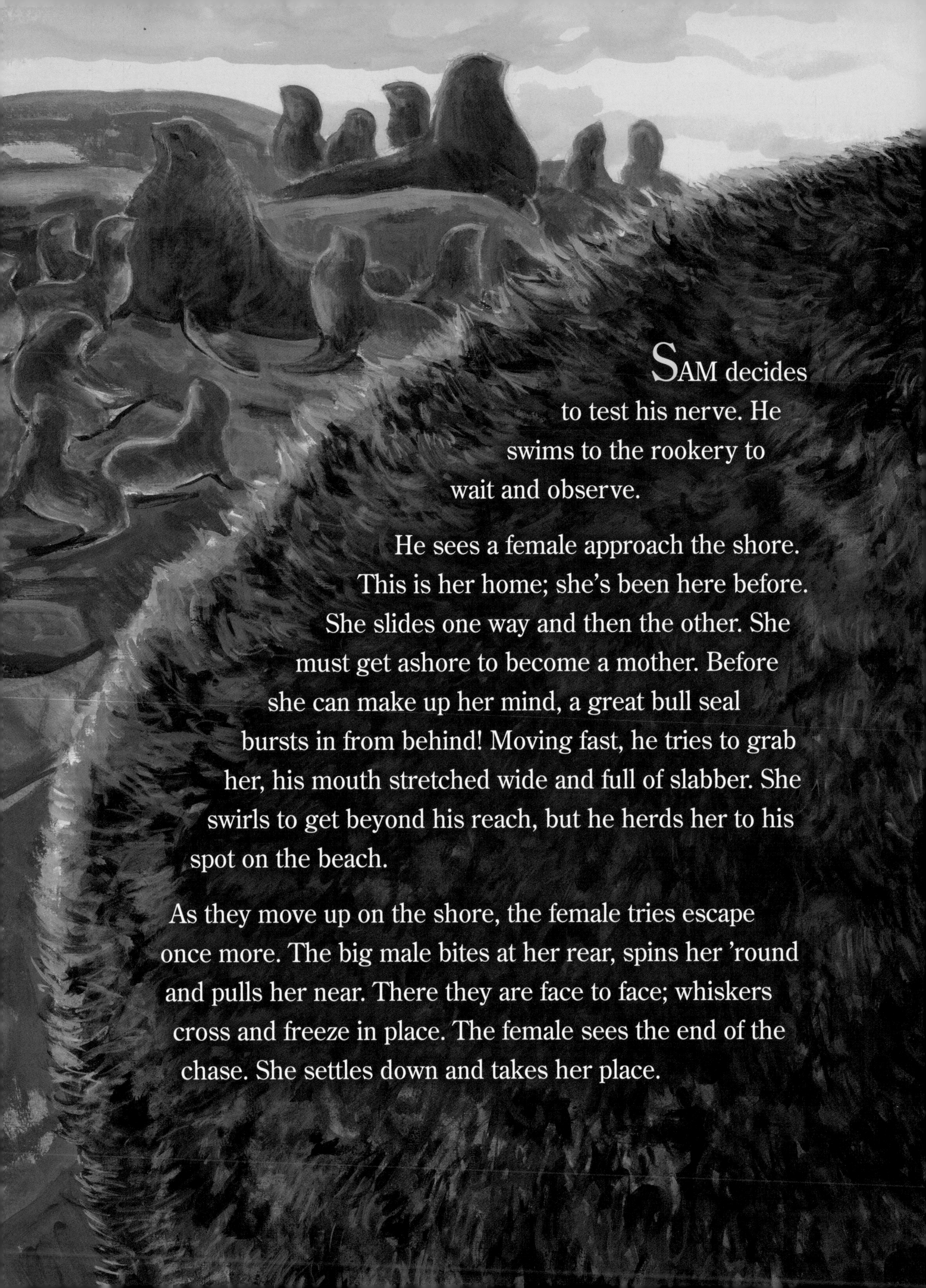

SAM decides to test his nerve. He swims to the rookery to wait and observe.

He sees a female approach the shore. This is her home; she's been here before. She slides one way and then the other. She must get ashore to become a mother. Before she can make up her mind, a great bull seal bursts in from behind! Moving fast, he tries to grab her, his mouth stretched wide and full of slabber. She swirls to get beyond his reach, but he herds her to his spot on the beach.

As they move up on the shore, the female tries escape once more. The big male bites at her rear, spins her 'round and pulls her near. There they are face to face; whiskers cross and freeze in place. The female sees the end of the chase. She settles down and takes her place.

The hulking male was pushy and eager. The sleek female was fierce but meager. While she was not sure of the male seal's worth, he did provide a place to give birth. Away to the side and all curled up, she gives birth to a slick, black pup. The little pup lifts his big wobbly head and soon decides it's time to be fed. The mother sniffs a coat smooth like silk. The pup sucks her sweet, rich milk. SAM scans the rocky ground, seeing other pups and mothers all 'round.

As SAM stares at the tranquil pairs, he hears a sound and turns around. In a flash of fur, fat, and foamy teeth, the big seal's speed is beyond belief. The huge beast reaches out to bite. The little seal swirls and flees in fright. The little seal plunges into the sea, twirling, whirling to get free.

SAM slips away through the calming ocean,
soothed by the sway of its gentle motion.

The small seal seems like a tiny whelp as he
weaves through cords of mighty kelp.

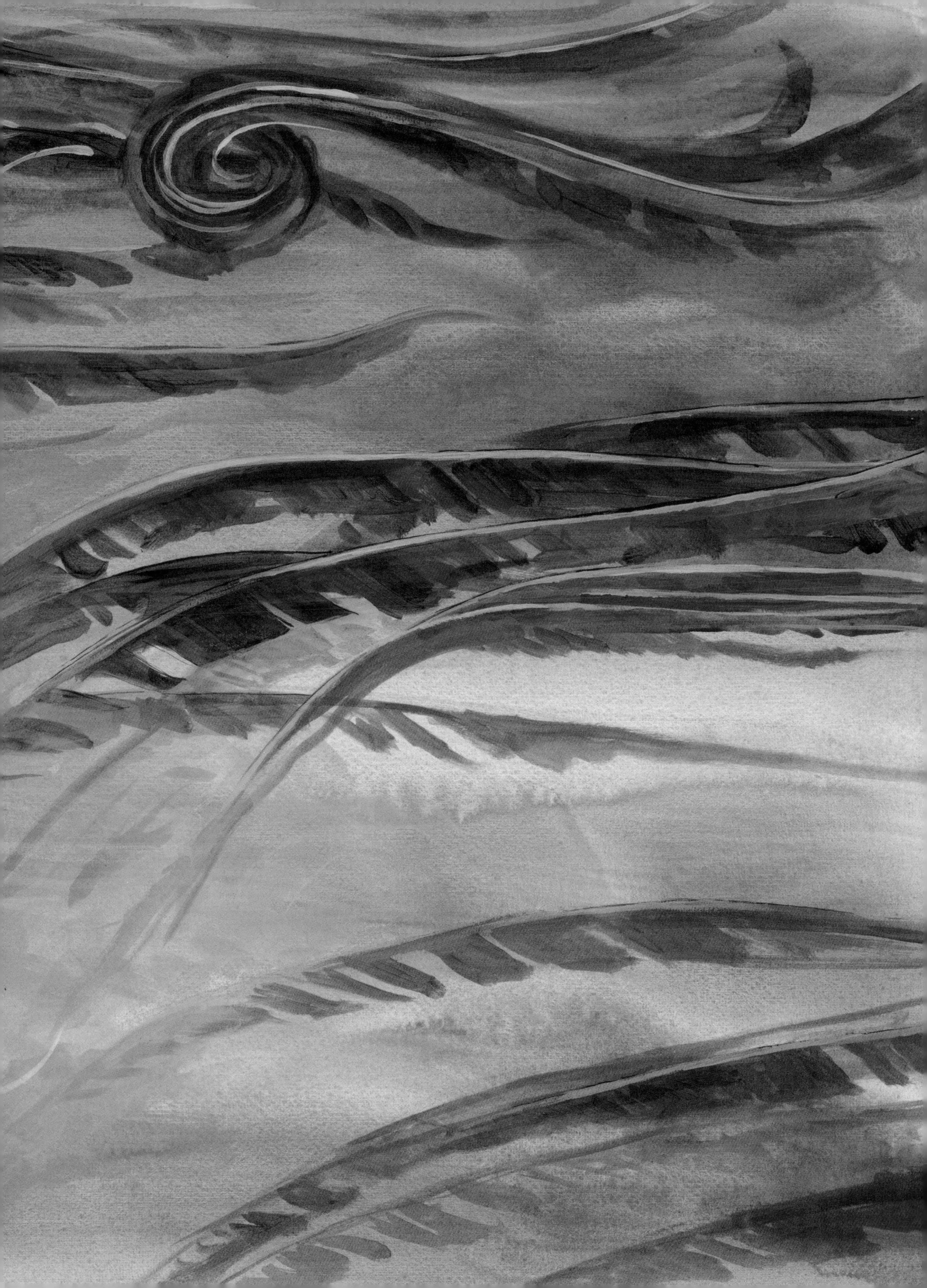

Hoping he has the size to get by on, SAM ends up spyin' a huge sea lion. Seeing the great beast so bulky up front makes SAM feel like a little runt. Part of the seal longs for the sea, but it is not the place he should be. Back on the shore he'll learn much more, so he turns around and heads for dry ground.

SAM goes back to the rookery on a whim. He sees pups splashing and thrashing, trying to swim.

As time moves on and days get warm, the rookeries begin to transform. The beachmasters haven't slept since who knows when. They haven't eaten for months on end. With all the fighting to defend their range, their bodies start to wither and change.

Females leave for weeks
to forage, building up
their blubber storage.
Pups gather together
making groups for play,
hoping their mothers will
return someday.

One pup hears a sound from the old sea bed. He calls back with a shake of his head. The plump, dark pup moves close to the sea, knowing what that sound might be. The big round pup calls like a lamb, looking to where his mother once swam. He climbs a rock and while he poses, his mother pops up and they touch noses!

Older pups now swim in playful groups. SAM sees them glide and turn in loops. Sam would love to stay and play, but it's time he's on his way. SAM will wander far from here, but he will return the very next year.

For ten winters SAM swims to far places. For ten summers, the same challenge he faces. Though his body changes and grows, pain and defeat are all that he knows. Each time SAM faces a beach defender, he's bitten and torn and forced to surrender.

SAM will never succeed coming from the ocean, but the little seal has another notion. He once saw a seal chased into the grassy waste, but away from the shore he's never been before. Beyond, beyond, far from the sea is the last place a seal should be. The biggest seals defend the shoreline, but what if he came in from behind?

Swimming through a stormy sea, SAM moves to the edge of the rookery. Across the rocks into the sand, he begins a slow crawl inland. Glancing across the strange new scenery, he plunges into the luscious greenery. Lumbering across the grassy ground, he stops to rest and look around.

Checking the shore with many a glance, the little seal sees his one big chance. SAM finally has an opportunity as old bag of bones lumbers back to the sea.

The empty spot within his sight, SAM slips in without a fight. Sometimes it helps to be the smartest and keenest if you can't be the biggest and meanest.

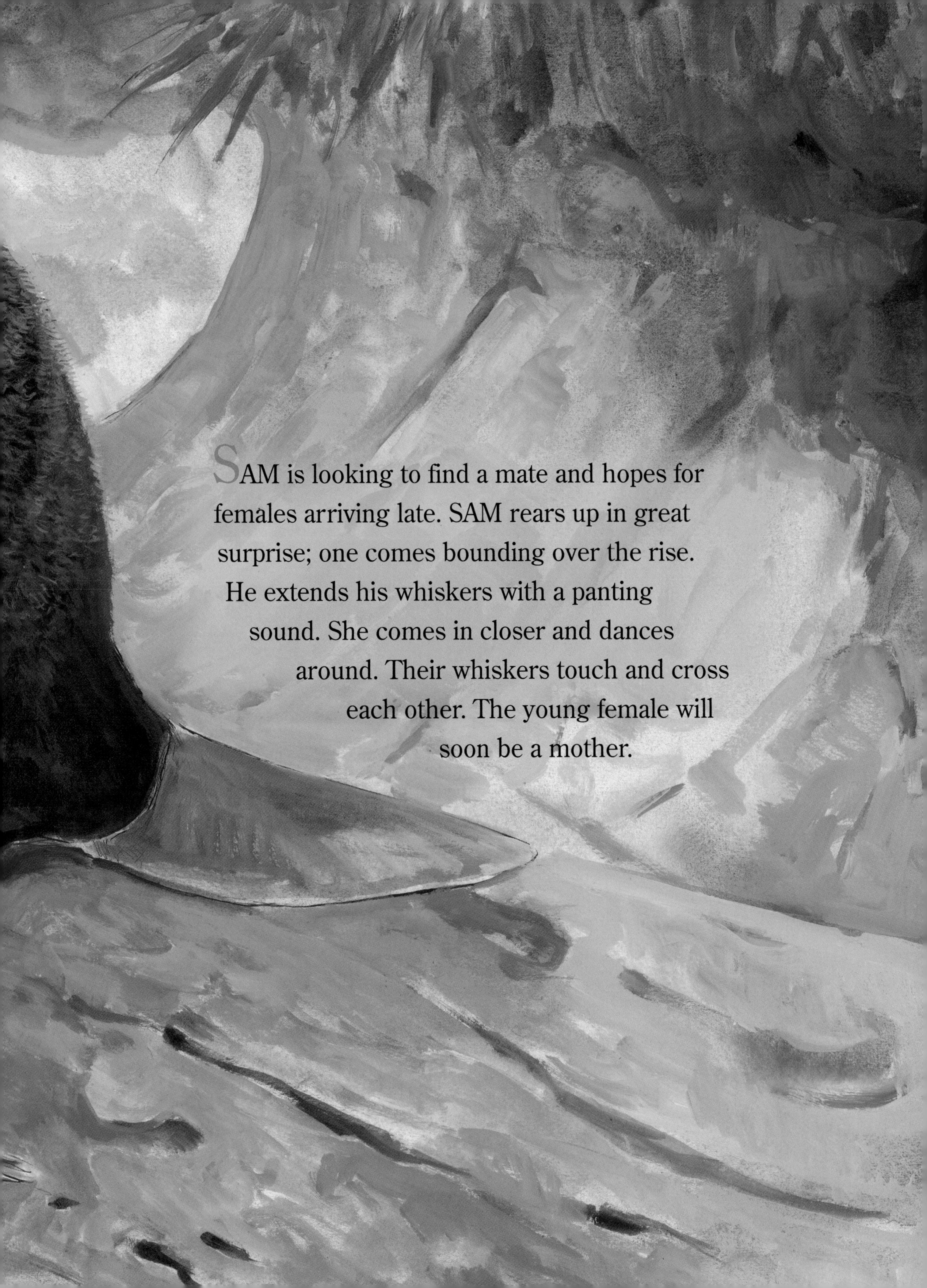

SAM is looking to find a mate and hopes for females arriving late. SAM rears up in great surprise; one comes bounding over the rise. He extends his whiskers with a panting sound. She comes in closer and dances around. Their whiskers touch and cross each other. The young female will soon be a mother.

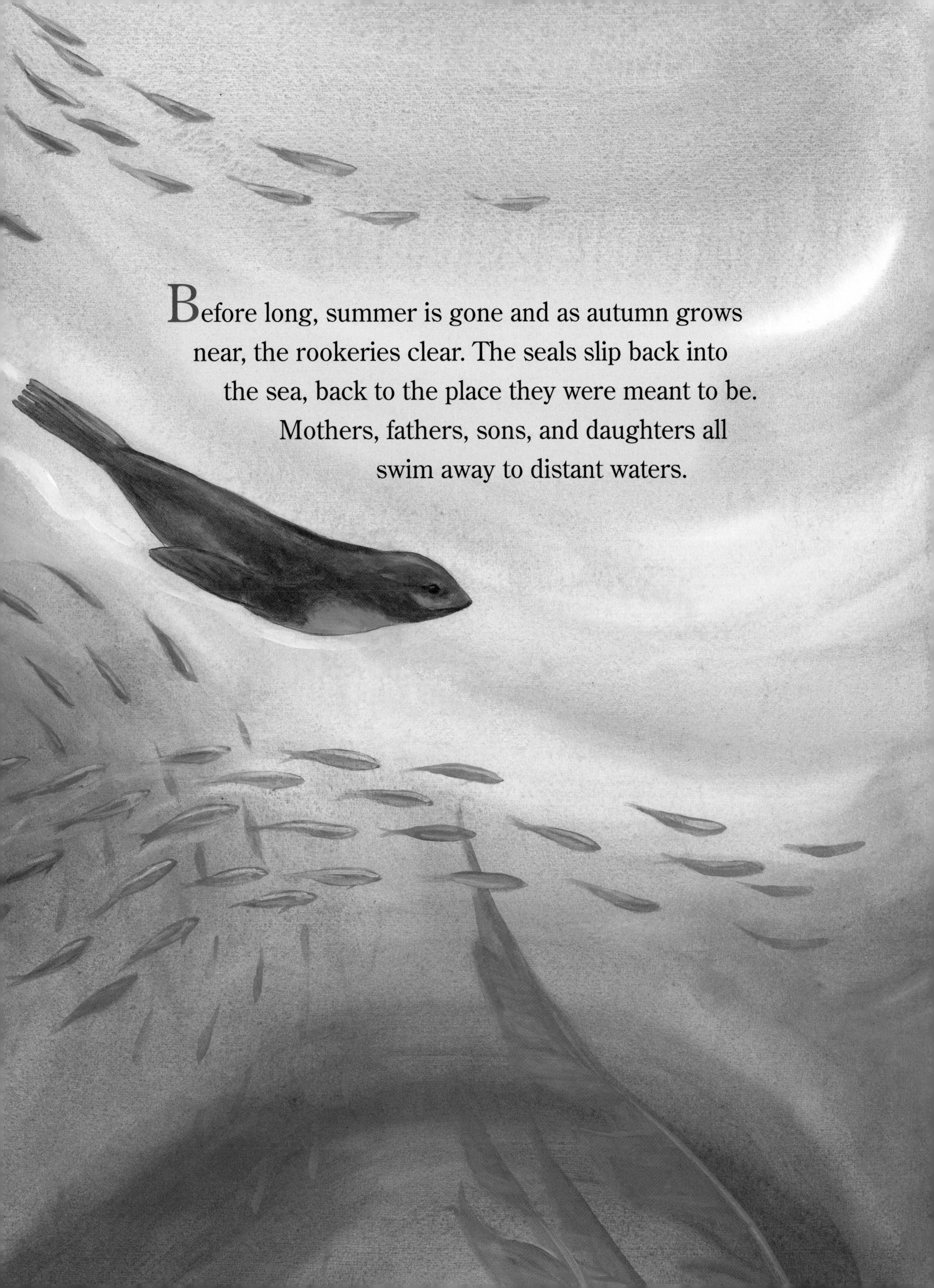

Before long, summer is gone and as autumn grows near, the rookeries clear. The seals slip back into the sea, back to the place they were meant to be. Mothers, fathers, sons, and daughters all swim away to distant waters.

Proceeds from the sale of this book go to support the Pribilof Islands Stewardship Program. Among other work, this program cleans beaches of marine debris and rescues fur seals that have become entangled.

The philosophy of the program is that the Pribilof Islands are our home; so is the world. We are interconnected, thus everything we do produces effects that we must acknowledge and be aware of now.

The vision of the program is to encourage protectorship, awareness, and responsibility for our home islands and the Bering Sea, which includes sharing these responsibilities with others.

THE NORTHERN FUR SEAL *(Callorhinus ursinus)* spends most of its life in the open ocean of the North Pacific, from California up through Alaska and down to Japan. They are only occasionally seen from land except when they gather on isolated rocky islands to give birth and breed. They travel hundreds of miles, farther than any other seal or sea lion, to reach their remote breeding grounds. Most fur seals go to the Pribilof Islands in Alaska's central Bering Sea, where, historically, several million fur seals converged annually. Their population in the Pribilofs in 2008 was less than one million and dropping rapidly.

During May and June of each year, the largest and strongest of the adult male northern fur seals, often called "beachmasters," arrive on land and fight for territories.

In late June and early July most of the pregnant females arrive. As they come ashore the beachmasters aggressively corral them into their territories. Within a day or two of arrival, the females give birth to their single pups. About five days after giving birth, females mate with a male territory holder, and then go to sea in search of food. Throughout the four-month period of pup rearing, females leave their young for one to two week periods while they go to sea to feed on fish and squid. Upon their return to land, they use sight, sound, and smell to find and nurse only their own pups, often among hundreds of others: a truly remarkable feat of recognition.

During the peak breeding season, young males are entirely excluded from the breeding areas or "rookeries" by the large and aggressive beachmasters. As a result, large groups of young males gather close to the rookery edges. These younger males are very curious and constantly probe the periphery of the rookeries, intently watching all that goes on. As July draws to a close, there are fewer new females and the big bulls become less aggressive, abandoning their territories and heading back to sea.

At first, the biggest males from the rookery edges fill and defend the openings, but as the season wears on the territory structure of the rookeries breaks down and the younger noncompetitive subadult males are able to gain access to what was once forbidden territory.

At the same time large "pods" of pups form as many females are away foraging. As the pups becoming increasingly mobile and learn to swim, they wander farther from their birth areas but still manage to reunite regularly with their mothers. In early November, about the time when the first snow begins to fly, pups and mothers separately leave the breeding areas. In the Pribilofs, they tend to swim south, through the Aleutian chain of islands and out into the open Pacific, until the following year when the cycle is repeated. Although much is known about this species, more than most animals, many questions remain unanswered, particularly about these early years at sea.

Dr. Stephen J. Insley
University of Victoria, Victoria, B.C.